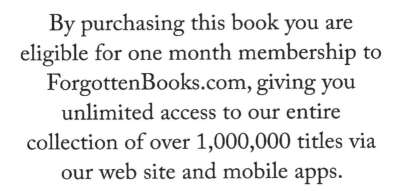

ISBN 978-0-265-85797-7
PIBN 10900276

NATIONAL BUREAU OF STANDARDS
Technical Note 307
Issued April 12, 1965

PHASE AND AMPLITUDE DIVERSITY IN OVER-WATER TRANSMISSIONS AT TWO MICROWAVE FREQUENCIES

H. B. Janes, A. W. Kirkpatrick
D. M. Waters and D. Smith
Central Radio Propagation Laboratory
National Bureau of Standards
Boulder, Colorado

NBS Technical Notes are designed to supplement the Bureau's regular publications program. They provide a means for making available scientific data that are of transient or limited interest. Technical Notes may be listed or referred to in the open literature.

For sale by the Superintendent of Documents, U.S. Government Printing Office
Washington D.C. 20402
Price: 40 cents

TABLE OF CONTENTS

PHASE AND AMPLITUDE DIVERSITY IN OVER-WATER
TRANSMISSIONS AT TWO MICROWAVE FREQUENCIES*

H. B. Janes, A. W. Kirkpatrick, D. M. Waters, and D. Smith

Propagation measurements at 9.4 and 9.2 Gc/s were
made over a 47 km, line-of-sight, over-water path at
Eleuthera Island in the Bahamas. The purpose was to
study the signal amplitude and phase variations at the two
frequencies, and in particular, the variations in the
phase difference and amplitude ratio ("differential amplitude") of the two frequencies. The results include power
spectra of phase and phase difference variations (from
about 10^{-4} to 4 c/s) and cumulative distributions of
amplitude, differential amplitude, and phase difference.

1. Introduction

A microwave signal transmitted over a fixed line-of-sight path
through the troposphere will exhibit time variations in both phase-of-
arrival and amplitude. Such variations are caused by fluctuations in
the three-dimensional structure of the atmospheric refractive index
along the radio path. Phase-of-arrival variations have been inves-
tigated in an extensive experimental program carried on by the
National Bureau of Standards using predominantly over-land paths
[Herbstreit and Thompson, 1956; Norton, et al., 1956; Thompson and
Janes, 1959; Thompson, Janes, and Kirkpatrick, 1960] . Although
these experiments were not designed to study amplitude variations,
the signals were observed to undergo occasional deep and prolonged
fading on paths of the order of 15 km in length. Two questions which
these previous experiments have left unanswered are: (1) how will
the signal phase and amplitude behave on an over-water path, and (2)
to what extent will the phase and amplitude variations be correlated in
two signals separated in frequency.

*This work was supported by the U.S: Air Force Missile Test Center
(MTDRM) under Air Force Order No. (08-606) 63-2147, Sept. 11, 1962.

An essential part of the Mistram precision missile trajectory measurement system being installed on the Atlantic Missile Range consists of microwave links which transmit phase information from the outlying terminals of two baselines to a central station where it is converted to target position and velocity information. Each link involves the simultaneous transmission of two signals in the neighborhood of 8 Gc/s separated by 256 Mc/s. The successful operation of the system requires that the phase and amplitude variations of one signal relative to the other remain below certain limits.

At the request of the Air Force Missile Test Center, the National Bureau of Standards in the fall of 1962 made a series of measurements at the Mistram installation on Eleuthera Island, British West Indies, to determine the magnitude of these relative phase and amplitude variations. The purpose of this report is to present the results of this work.

2. General Description of the Experiment

2.1. Path and Equipment Location

The baseline microwave link chosen for the measurements at Eleuthera extends from the Mistram central site near the Auxiliary Air Force Base in a southerly direction to the Powell Point site at the southwestern tip of the island, a distance of 47.32 km (figure 1). Except for the immediate foreground at each end, the entire path is over water. The antenna at the central site terminal is a 3-meter parabolic dish mounted on a building roof 4.1 meters above local ground and 32 meters above mean sea level. At the Powell Point terminal, a 2.4-meter parabolic antenna is mounted on a guyed steel tower at a height of 59 meters above local ground and 71 meters above mean sea level.

The klystrons and mixers were placed as close as possible to each antenna, and coaxial cable was used to carry the IF signals from that point on. (At Powell Point, the mixing took place directly behind the antenna; at the Central site, the mixer was at the end of about 20 m of waveguide, almost all of which was inside an air-conditioned building). Consequently, the phase and amplitude variations discussed here are only those contributed by the propagation medium and by any mechanical motion of the two Mistram antennas involved. A description of the phase and amplitude measurement equipment is given in the appendix.

Meteorological measurement equipment consisted of two micro-wave refractometer sensors mounted on the antenna tower at Powell Point at heights of 6 and 43 meters above ground (18 and 55 meters above mean sea level). Another refractometer sensor was mounted on a mast 5.5 meters above ground (33.5 meters above m. s. l.) at the Central site. In addition, wet and dry bulb thermometers and aneroid barometers were placed in shelters 1.5 meters above ground at each terminal. Anemometers to record horizontal wind velocity were located at each terminal, 8 meters above ground at the Central site and 13 meters above ground at Powell Point.

2.2. Data Output

Experimental data were recorded during three periods, each approximately two days in length. In general, they consisted of time variations in the following parameters:

1. The phase-of-arrival of a 9.2 and a 9.4 Gc/s signal transmitted over the 47 km path, each phase record, in effect, being relative to an arbitrary phase reference at 9.2 and 9.4 Gc/s, respectively. Hereafter, these will be referred to as the "single frequency phase" or ϕ data. These variables were each split into two recordings, a low-pass recording with a pass band from 0 to 10 c/s, and a band-pass recording with a pass band from about 0.25 to 10 c/s. The low-pass recording circuits were equipped with an automatic scale-shift feature which shifts the variable in precise increments of 180 degrees whenever it threatens to exceed the scale limits of the recorder. The resulting records combine relatively high resolution with long-term continuity.

2. The difference between the phase-of-arrival of the 9.2 and 9.4 Gc/s signals, hereafter referred to as "phase difference" or $\Delta \phi$ data. This variable was also split into a low-pass recording (from 0 to 5 c/s) and a band-pass recording (from .04 to 15 c/s). Although it was not possible to calibrate this record in terms of absolute phase difference, the time variations in phase difference were recorded relative to an arbitrary reference in the same way as were the single frequency phase variables.

3. The amplitudes of the 9.2 and 9.4 Gc/s signals. These records have a pass band extending from 0 to about 15 c/s. Since the purpose of this part of the experiment was to study amplitude changes at one frequency relative to those occurring on the other, no attempt was made to calibrate these data in absolute units of received signal power. Hence, they are discussed here in terms of decibels relative to an arbitrary zero.

4. Refractivity variations at the low-level refractometers at both terminals, and the variations in refractivity difference between the refractometers located 6 and 43 meters above ground at Powell Point.

5. Wind speed and direction at each terminal.

6. Wet and dry bulb temperature and barometric pressure at each terminal.

The phase and amplitude data were recorded continuously in analog form on magnetic tape, with occasional real-time monitoring on paper strip charts. The wind speed and direction and refractivity were recorded on relatively slow-moving paper strip charts to indicate long-term changes. The temperature and pressure were logged at half-hour intervals throughout each recording period.

3. Phase Variation Data

3.1. General

In general, the single-frequency phase (ϕ) data were very similar to those obtained in previous experiments. The gross, hour-to-hour and day-to-day variations caused by long-term changes in the mean refractive index along the path covered many hundreds of phase degrees. This is illustrated in figures 2, 3, and 4, which show the 9.4 and 9.2 Gc/s phase records sampled at 30-minute intervals and plotted versus time. The phase ordinate here is expressed in terms of changes in the apparent radio path length both in parts per million and in wavelengths at the two radio frequencies (i.e., 1λ = 360 phase degrees). (Only the first half of the second run, October 30 - November 1, is plotted here. Equipment trouble at the middle of the run resulted in loss of continuity, and hence precluded showing the long-term variations for the entire run. However, the spectrum and distribution analysis described below contain samples taken over the entire recording period.)

The corresponding long-term changes in surface refractivity* measured at both terminals along with the vertical gradient variations at Powell Point are plotted in figures 5, 6, and 7.

Some correlation between the grossest features of the phase and refractivity can be seen, but in general, the correlation is poorer than in the case of shorter, over-land paths used previously [Thompson and Janes, 1959; Thompson, Janes, and Kirkpatrick, 1960] . This is not surprising, since the refractivity data were taken on land and were influenced heavily by local terrain, vegetation, etc. Hence they do not accurately represent the mean refractivity along the over-water path. The long-term variations in the vertical refractivity gradient at Powell Point are plotted in the lower part of these graphs. Similar long-term plots of the phase difference ($\Delta \phi$) data are shown in figures 8, 9, and 10.

In interpreting these data, it is helpful to consider a highly over-simplified model of the propagation path and medium. In this model, the received radio energy is confined to an infinitely thin ray path between the two antennas with no multipath permitted. At any given instant the phase of the received signal at radio frequency f relative to that emerging from the transmitting antenna would be given by:

$$\phi_f(t) = \frac{360 f L[1 + \overline{N}(t) \times 10^{-6}]}{C}$$

---

*The relationship between the refractivity, N, and the refractive index, n, is given by:

$$N = (n - 1) \times 10^6$$

In other words, a change of 1 N-unit represents a change of about 1 part per million in n.

where

$$\phi_f(t) \;=\; \text{phase-of-arrival in degrees}$$

$$\overline{N}(t) \;=\; \text{refractivity averaged over the line path}$$

$$L \;=\; \text{path length}$$

$$C \;=\; \text{velocity of propagation in a vacuum}$$

$\phi_f(t)$ then would be a linear function of $\overline{N}(t)$, and the derivative

$$\frac{d\phi}{d\overline{N}}$$

would be proportional to f. Similarly, the phase difference between two signals at f_1 and f_2 is given by

$$\Delta\phi(t) \;=\; \phi_{f_1}(t) - \phi_{f_2}(t) \;=\; \frac{360(f_1 - f_2)\, L(1 + \overline{N} \times 10^{-6})}{C}$$

and the derivative $\dfrac{d\Delta\phi}{d\overline{N}}$ would be proportional to $f_1 - f_2$. As $\overline{N}(t)$

varies with time, $\phi_{f_1}(t)$, $\phi_{f_2}(t)$ and $\Delta\phi(t)$ would (under this model) be

perfectly correlated with $\overline{N}(t)$, and the variations in $\Delta\phi$ would be

smaller than those in ϕ_{f_1} by the ratio

$$\frac{f_1 - f_2}{f_1}$$

Even in this simple model, it would be very difficult to determine $\overline{N}(t)$ so as to provide a corrector for the ϕ and $\Delta\phi$ variations. (The N data taken at the path terminals are obviously a poor estimate of $\overline{N}$ in this case.) However, it would be possible to use a single frequency phase record to compensate for the correlated part of the phase difference variations.

The observed data behave somewhat in the manner predicted by the use of this model, but they also differ from it in some important respects. For example, the long-term variations in single-frequency phase, ϕ, at 9.2 and 9.4 Gc/s shown in figures 2, 3, and 4, are obviously well correlated with each other. Furthermore, the corresponding variations in phase difference between the 9.4 and 9.2 Gc/s signals (i. e., $\Delta \phi$) are also well-correlated with either of the single frequency phase records, as can be seen by comparison of these figures with figures 8, 9, and 10. The lower graph in each of the latter figures is simply

$$\Delta \phi_R (t) = \Delta \phi(t) - \frac{f_1 - f_2}{f_1} \phi_{f_1}(t).$$

It shows the extent to which the effects of long-term changes in $\Delta \phi$ could be reduced by subtracting from them the scaled-down ϕ variations and hence removing the $\Delta \phi$ variations caused by the difference in radio frequencies.

It should be pointed out that

$$\Delta \phi_R (t) = \frac{f_2}{f_1} \phi_{f_1} (t) - \phi_{f_2} (t)$$

and hence is directly proportional to the difference in radio path lengths that would be observed at f_1 and f_2.

3.2. Power Spectra of Phase Data

The relationships among these phase and phase difference data are illustrated in a more systematic way in the power density spectra of single frequency phase and phase difference shown in figures 11, 12, and

13. (In each figure, the lower frequency portion was computed from the data for the entire run, digitized at one-minute intervals; the higher frequency portion is the median of 12 to 15 individual spectra computed from three-minute data samples digitized at the rate of 30 points per second.) The purpose of the power spectrum is to indicate how much of the mean square fluctuation (variance) of a given variable is contributed by a particular band of fluctuation frequencies. Since the ordinate is given in units of variance (degrees squared) per unit bandwidth, the variance contributed by any band of spectral frequencies can be computed simply by integrating that portion of the spectral density curve.

Of particular interest here are the relative magnitudes of the ϕ and $\Delta\phi$ spectra. At the lowest fluctuation frequencies (below 10^{-4} c/s), the phase difference, $\Delta\phi$, is approximately three orders of magnitude smaller in spectral density (i.e., variance) than the single frequency phase, ϕ. This is in accord with the simplified picture given above, i.e.,

$$\text{variance of } \Delta\phi = \left(\frac{f_1 - f_2}{f_1}\right)^2 \times [\text{variance of } \phi_{f_1}]$$

where

$$\left(\frac{f_1 - f_2}{f_1}\right)^2 = \left(\frac{9.4 - 9.2}{9.4}\right)^2 = 4.5 \times 10^{-4}.$$

However, the $\Delta\phi$ spectrum increases relative to the ϕ spectrum as the spectral (fluctuation) frequency increases until the two spectra cross in the neighborhood of 1 c/s. Beyond 1 c/s, the $\Delta\phi$ spectrum tends to assume a value twice that of the ϕ spectrum as the fluctuation frequency increases.

From these spectra and the above simple model we can conclude the following: at the lowest spectral frequencies, ϕ_{f_1} and ϕ_{f_2} are so well correlated that the principle difference between them is the slight difference in magnitude caused by the difference in radio frequencies. Consequently their difference, $\Delta\phi$, is well-correlated with both ϕ_{f_1} and ϕ_{f_2}, but reduced in magnitude by

With increasing spectral frequency, the correlation of ϕ_{f_1} and ϕ_{f_2} reduces, causing a relative increase in $\Delta\phi$ fluctuations. Above about 1 c/s, the correlation of ϕ_{f_1} and ϕ_{f_2} is apparently zero, since the spectral density of $\Delta\phi$ is approximately twice the spectral density of ϕ_{f_1} or ϕ_{f_2}. At this point it might be helpful to recall the following relationship:

$$\text{var } \Delta\phi = \text{var } \phi_{f_1} + \text{var } \phi_{f_2} - 2r \sqrt{\text{var } \phi_{f_1} \text{ var } \phi_{f_2}}$$

where "var" indicates a variance or mean squared fluctuation, and
r = correlation coefficient between ϕ_{f_1} and ϕ_{f_2}

The fact ϕ_{f_1} and ϕ_{f_2} do not remain correlated throughout this range of fluctuation frequencies suggests that the simplified single ray path model described above is not very realisitic. Actually, the radio energy sent between the Mistram Central Site and Powell Point travels over many slightly differing paths simultaneously. These include sea reflections as well as refraction and scattering from atmospheric irregularities. According to scatter propagation theory [Wheelon, 1959], the amplitude of the energy scattered in a given direction from an atmospheric irregularity is a function of (among other things) the radio frequency. The received signals, then, are the vector sum of a number of contributing sources, and the resultant phase of the signal at f_1 contains some rapidly fluctuating components due to scattering which are statistically independent of the corresponding phase fluctuations at f_2.

An interesting sidelight of the spectrum analysis is the appearance of two pronounced spikes in the ϕ spectrum during the run of October 30 to November 1 (figure 12). These appear at about 0.55 and 2.4 c/s and are apparently caused by small resonant vibration of the antenna tower at Powell Point. The wind velocity data for each run are plotted in figures 14, 15, and 16, and it should be noted that the winds were at their maximum at Powell Point during this run.

3.3. Comparison of Power Spectra with Theory

E. C. Barrows [1964] of CRPL has computed theoretical power spectra of ϕ and $\Delta \phi_R$ and these are shown in figure 13a along with a derived spectrum of $\Delta \phi$ and corresponding experimental spectra of ϕ and $\Delta \phi$ from the November 2-4 run. ($\Delta \phi_R$ was not recorded directly in this experiment and hence an experimental $\Delta \phi_R$ spectrum could not be obtained. However, figures 8, 9, and 10 show a rough estimate of the long term variations of this parameter obtained from the ϕ data as described in Section 3.1.) The theoretical ϕ and $\Delta \phi_R$ spectra were computed from forward scatter theory using the appropriate path length and antenna parameters and assuming (a) a mean cross-path wind velocity component of 2 m/sec; (b) Taylor's hypothesis relating time and space fluctuations in the turbulent atmosphere; (c) the absence of surface-reflected multipath; and (d) a slope of f^{-2} for the ϕ spectrum (chosen as a convenient approximation of the observed slopes). The theoretical spectrum of $\Delta \phi$ was produced by first deducing the co-spectrum of ϕ_1 and ϕ_2 from the relative magnitudes of the theoretical ϕ and $\Delta \phi_R$ spectra and then using this cospectral function to compute the $\Delta \phi$ spectrum.

There appears to be good general agreement between the experimental and theoretical ϕ and $\Delta \phi$ spectra (the placement of the theoretical spectra on the vertical scale is arbitrary). The agreement could probably be improved by assuming a steeper slope for the ϕ spectrum, e.g., $f^{-2.3}$ instead of f^{-2}. This agreement is somewhat surprising in view of the fact that the theory assumed no multipath (for the sake of simplicity). Since the path geometry offers ample opportunity for strong sea multipath effects, (as evidenced by the fading data discussed below) the influence of multipath on the ϕ and $\Delta \phi$ spectra in this range of spectral frequencies may not be great.

One interesting discrepancy between theory and experiment occurs at the high-frequency end of the spectra. The theoretical ϕ spectrum shows a decided steepening in slope above about 0.8 c/s caused by antenna aperture smoothing. The corresponding experimental curve shows some steepening but fails to follow the theoretical curve at the highest frequencies plotted. The higher spectral density of the data here may be due to measurement and data processing noise. The latter dominated the spectrum above 5 c/s and may have influenced the spectrum somewhat between 1 and 5 c/s.

3.4. Cumulative Distributions of Phase Data

In order to provide an estimate of the probability that the variations in phase difference will exceed a given range during a given time period, cumulative distributions were obtained for each of the three-minute samples used in the spectrum analysis described above, and also for 30-minute periods spaced as nearly as possible at 4-hour intervals throughout each run.

The three-minute distributions are shown plotted on Gaussian probability paper in figures 17 through 22. (In these and subsequent distribution graphs, the times noted refer to the beginning of the sample.) Of the 35 samples shown, 34 have 1-99% ranges of 5 degrees or more, while 25 have 5-95% ranges of 5 degrees or more. It should be borne in mind that these distributions are taken from a band-pass recording with lower cutoff at 0.04 c/s, and consequently, the ranges of variation shown here are systematically somewhat smaller than the actual ranges that would have been obtained in the absence of filtering.

To illustrate the general appearance of the short-term variations in phase difference, two examples of chart records are shown in figure 23. They represent the "quietest" and "noisiest" of the band-pass samples used in the distribution analysis. In figure 24, two examples of longer-term variations are shown. These are 30 minute samples transcribed from the low-pass record and were chosen primarily to coincide with two of the three differential fading samples shown in figure 28. It should be noted from these samples that phase difference changes on the order of 20 degrees occurred within a period of one second, and that changes twice that size occurred within a minute. (The very large sweeps of the pen extending over nearly the width of the chart in the 1813-1843 sample in figure 24 are caused by the automatic 180° scale-shift feature employed in all the low-pass phase recording circuits).

The thirty-minute cumulative distributions are shown in figures 25, 26, and 27. They display a wide variety of slopes, with 1-99% ranges extending from less than 2 degrees to more than 60 degrees. It should be pointed out here that this is not an unbiased random sample of phase difference data. As will be explained in the next section, the signal levels sometimes went below the operating threshold of the equipment during periods of deep and violent fading, resulting in loss of both phase and amplitude data. Since the phase difference variations were often observed to be unusually large during periods of weak signal, it is probable that these sample distributions are somewhat biased in that they fail to represent the worst propagation conditions encountered.

-11-

4. Signal Amplitude Data

The time variations in the amplitude of the 9.4 and 9.2 Gc/s signals were recorded continuously on magnetic tape. The differential amplitude, or fading of one signal relative to the other, was subsequently generated in analog form in the laboratory.

Three examples of chart transcriptions of the 9.4 Gc/s amplitude function along with the corresponding differential amplitude are shown in figure 28. It will be seen at once that the amplitude data are characterized by long periods of strong signals with negligible fading, interrupted by periods of significant fading lasting from several minutes to several hours. The long-term fading characteristics of the 9.4 Gc/s signal for each run are shown in figure 29. Here the medians of half-hour periods are plotted versus time, along with vertical bars indicating the 10% to 90% fading ranges taken from cumulative distributions of selected half-hour periods. In the lower portion of each graph, the 10-90% fading range of the differential amplitude is shown for the same half-hour periods. The cumulative distributions of amplitude and differential amplitude for the selected periods are shown in figures 30 through 41. To illustrate the wide variety of fading conditions represented in these cumulative distributions, figure 42 shows transcriptions of the differential amplitude variations for three of the periods. They were chosen to represent both quiet and active fading periods. For example, the 0230 to 0300 sample shows a 10 db fade of one signal relative to the other occurring within one minute, while the fading during the entire 1727 to 1757 sample covered only about 0.2 db. The 1813 to 1843 sample is especially interesting because it contains both quiet and active differential fading within a relatively short period of time. Comparing it to the phase difference transcription for the same period in figure 24, there was apparently good time correlation between the magnitude of the differential amplitude and phase difference variations. However, with the possible exception of the deep fading periods, some of which were necessarily excluded from this analysis, this was not true in general, i.e., the observed variances of phase difference and differential amplitude were found to be essentially uncorrelated.

From figures 28 and 29 and from the sample data transcriptions, it can be seen that the ranges of short-term (i.e., minute-to-minute) variations in both amplitude and differential amplitude are very small (on the order of 1 db) during strong signal periods. A general decrease in signal strength was invariably accompanied by a pronounced increase in the fading range of both signals and of their decibel difference. The

-12-

periods of weak signal, then, are of primary interest in the present study. The minimum signal level at which the phase and amplitude measurement equipment will operate was approximately 20 db below the maximum signal strength observed. When the signal dropped below this threshold, both phase and amplitude data were lost for periods of a few seconds to a few minutes. Unfortunately, this occurred occasionally during all three runs, with the results that the fading ranges and cumulative distributions shown here do not include some of the worst fading conditions. (About 13% of all half-hour periods contained at least momentary losses of data believed to be caused by deep fading.) The half-hour samples for the distribution analysis were spaced as nearly as possible at 4-hour intervals, but considerable deviation from this rule was sometimes. required to avoid these occasional sections of lost data. The bias that would inevitably result from shunning periods of lost data (i.e., deepest fading) was partially compensated for by the arbitrary inclusion of three or four additional signal amplitude samples from each run, deliberately chosen to represent periods of deep fading (but still avoiding periods of lost data, of course). This compensation may have been reasonably successful, since 15% of the resulting samples have peak-to-peak differential fading ranges in excess of 10 db, while visual inspection of the differential fading records for all three runs indicates 17% of all half-hour periods have peak-to-peak ranges greater than 10 db.

The fading periods do not appear to be correlated with the refractive index data (figures 5, 6, 7), wind velocity (figures 14, 15, 16), phase data, tidal fluctuations (figure 43), or time of day. (However, some of the fading periods did coincide with the passage of rain showers across the path.) The most readily-suspected cause of this deep fading is sea multipath, i.e., the interference of direct and strong sea-reflected signal components. The phase relationship of the two components could easily be changed from reinforcement to cancellation by a very moderate change in the refractive index gradient between the sea surface and the minimum height of the direct ray. If this were the case, one might expect some correlation between the fading and the refractive index data in figures 5, 6, and 7. However, the latter data were obtained at the path terminals, several hundred meters from the shore in both cases, and hence should not be expected to be closely correlated with refractive index gradient variations over the sea surface.

Another possible (but in this case less likely) cause of fading is one described by Dutton [1961] which requires a vertical refractive index profile such as would produce a low level, surface-based radio "duct". If the upper limit of this duct lies above one path terminal and below the other, it is possible that the radio waves will be refracted in such a way that the receiving terminal will lie in a "shadow zone" with respect to the transmitter, with resulting deep fading.

Simultaneous measurements of signal amplitude on this path and the refractive index structure (on the path rather than at the land terminals) from the sea surface to a height of about 100 meters would be very valuable in studying the propagation phenomena involved.

5. Sea Reflection Screen

Since it appeared likely that reflections from the sea surface would have an important influence on phase and amplitude variations, a screen was designed and constructed by Pan American World Airways to be placed in the water at the nominal reflection point to partially block off sea reflections. (Bussey [1950] had successfully used such a screen on a much shorter path.) The screen actually consisted of two sections mounted in the same vertical plane with a 6 meter horizontal separation between them. Each section was a square 6 meters on a side, covered with galvanized hardware cloth having a grid spacing of 0.5 cm.

The screen was put into position at about noon on November 2, and remained there throughout the November 2-4 run. Although the data from this run contain unusually long periods during which the signals exhibited very little fading, they also contain periods during which some of the deepest and most rapid fading occurred. Keeping in mind the usual differences in data gathered in successive runs under "identical conditions" (i.e., nothing permitted to change but the weather!), one cannot say that the data reveal any evidence that the screen produced a significant effect in the data. Apparently, if such effects exist, they are sufficiently subtle to require a more extensive experimental study to detect them.

6. Conclusions

The conclusions drawn from the results of this experiment can be summarized as follows:

6.1. Phase Variations

Although the phase variations in the 9.4 and 9.2 Gc/s signals are well-correlated at low fluctuation frequencies, the correlation deteriorates toward the high frequency end of the power spectrum and is essentially zero above about 1 cycle per second. Hence virtually all of the phase difference variation at low spectral frequencies is attributable to the difference in radio frequencies and is well-correlated with both the 9.4 and 9.2 Gc/s phase variables. However, the loss of correlation at higher spectral frequencies is indicated by the relative increase in the phase difference spectral density at the higher frequencies.

A cumulative distribution analysis of the phase data indicates that in 34 out of 35 cases the phase difference varied by more than 5 degrees during a 3-minute period, and that changes in phase difference on the order of 20 degrees occurred within a period of 1 second.

The variance of phase difference showed no correlation with time of day, tide, or atmospheric refractive index and wind data gathered at the path terminals.

6.2. Signal Amplitude Variations

A wide variety of fading conditions were observed, in terms of both the variations in signal amplitude at either frequency and the variations in the differential amplitude. The data indicate that a 10 db change in differential amplitude can occur within a period of one minute. They also indicate a 0.17 probability that during a given half-hour period, the differential amplitude of the two signals will vary by more than 10 db.

As in the case of the phase data, the differential fading range showed no correlation with time of day, tide, wind velocity or refractive index data.

A specially-designed screen placed at the nominal sea reflection point in an effort to eliminate multipath interference from the sea surface failed to produce any clearly identifiable effect in the fading data.

7. References

Barrows, E. C. (December, 1964), Private Communication.

Bussey, Howard E. (Dec. 1950), Reflected ray suppression, Proc IRE 38, No. 12, 1453.

Dutton, E. J. (June 16, 1961), On the climatology of ground-based radio ducts and associated fading regions, NBS Tech. Note 96.

Herbstreit, J. W., and M. C. Thompson, Jr. (July 1956), Measurements of the phase of signals received over transmission paths with electrical lengths varying as a result of atmospheric turbulence, IRE Trans. AP-4, 352-358.

Norton, K. A., J. W. Herbstreit, H. B. Janes, K. O. Hornberg, C. F. Peterson, A. F. Barghausen, W. E. Johnson, P. I. Wells, M. C. Thompson, Jr., M. J. Vetter, and A. W. Kirkpatrick (1956), An experimental study of phase variations in line-of-sight microwave transmissions, NBS Monograph 33.

Thompson, M. C., Jr., and H. B. Janes (Feb. 1959), Measurements of phase stability over a low-level tropospheric path, J. Res. NBS 63D (Radio Prop.), No. 1, 45-51.

Thompson, M. C., Jr., H. B. Janes, and A. W. Kirkpatrick (Jan. 1960) An analysis of time variations in tropospheric refractive index and apparent radio path length, J. Geophys. Res. 65, No. 1, 193-201.

Wheelon, Albert D. (Apr. 1959), Radio-wave scattering by tropospheric irregularities, J. Res. NBS 63D (Radio Prop.), No. 2, 205-233.

8. Appendix

8.1. Phase and Amplitude Measurement Equipment

The equipment used for two-frequency phase and amplitude measurements in this experiment can be described briefly as follows: The outputs of two frequency-stabilized c-w klystrons are combined, passed through a circulator and onto an antenna. The signal received at the remote antenna is divided between two transponders; each transponder retransmits its received frequency with a small offset in frequency. The transponder signals are received at the first antenna and passed through the circulator into a mixer where they are mixed with a small portion of the output of the stabilized klystrons. The output of the mixer is amplified by an IF amplifier for each transmitter transponder pair. The output of each IF is sent to a phase meter and amplitude detector. The IF signals produced at the transponder are sent to the control end by a 4.9 Gc/s FM link where they are used as phase references for the phase meters. The IF signals are fed into mixers and the difference frequencies drive a phase meter to provide the phase difference information. The IF signals from the transmitter end provide the amplitude variation information.

Referring to the block diagram in figure 44, a 10 Mc/s secondary frequency standard is used as a frequency reference for stabilizing the klystrons at the control end of the path. The oscillator stays within a few parts in 10^9 over a wide temperature range and ages at a rate of 3 parts in 10^9 per day. A times 20 multiplier raises the frequency to 200 Mc/s and the power level to about 0.5 watt. The 200 Mc/s is fed to the harmonic generators. A 20 db crossguide coupler supplies L.O. power from the 9.2 Gc/s klystron (K_1) to a mixer $(X_1$, 1N23D); the output of the harmonic generator (H_1) goes through a ferrite isolator to the mixer. The output of the mixer goes to a 500 kc/s phase-lock circuit which locks the klystron ±500 kc/s away from the 9.2 Gc/s output of the harmonic generator.

The 9.4 Gc/s klystron is locked ±93 kc/s away from the 9.4 Gc/s output of the harmonic generator (H_2) by the same technique. A hybrid tee is used to combine the output of the two klystrons and provide some isolation between them. Ferrite isolators are used between each klystron and the hybrid tee to provide additional isolation. The combined output goes to the antennas through a ferrite circulator; the circulator isolates the incoming from the outgoing signal on the antenna. A mixer (X_3) mixes

the incoming signal with a small portion of the klystron outputs and provides the difference signals between the outgoing and incoming signals.
At the other end of the path, the signals coming off of the antenna are
split by a hybrid tee; each output of the hybrid tee goes through a circulator (C_2 and C_3) to a mixer (X_{1a} and X_{2a}). The 9.2 Gc/s klystron
(K_{1a}) is phase-locked to the incoming 9.2 Gc/s signal ± 50 kc/s by
way of C_2 and X_{1a} and the 9.4 Gc/s klystron (K_{2a}) is phase-locked to
the incoming 9.4 Gc/s signal ± 80 kc/s by way of C_3 and X_{2a}. Because
of the means used to take the phase difference, the klystrons must be
both either higher or lower in frequency than their incoming signal.
The outputs of the klystrons (K_{1a} and K_{2a}) are added in the hybrid tee
and go to the antenna. The 80 kc/s and 50 kc/s signals are amplified
in narrow band amplifiers and multiplexed on to a 4.9 Gc/s FM transmitter. The 4.9 Gc/s FM receiver at the control end separates the
80 kc/s and 50 kc/s signals. The output of mixer X_3 at the control end
goes to 50 kc/s and 80 kc/s narrow-band amplifiers. The two 50 kc/s
signals go to a phase meter which provides the 9.2 Gc/s range variation
measurement, and the 80 kc/s signals go to a phase meter to provide
the 9.4 range variation measurement. The 50 kc/s and 80 kc/s signals
from the control end go to a mixer, and the two signals from the FM
receiver go to a mixer; the two 30 kc/s difference signals go to a third
phase meter which provides the range difference variation measurement.
Because the 9.4 and 9.2 Gc/s signals are each transmitted over the
path twice (once in each direction), the phase variations actually measured
are twice as large as would be the case if the path were only traversed
once and the incoming phase were compared to a local phase reference.
In this report, all phase data have been reduced by the factor of two to
represent "one-way" phase variations. The output of the 50 kc/s and
80 kc/s narrow band amplifier at the control end goes to amplitude
detectors for signal amplitude variation measurements.

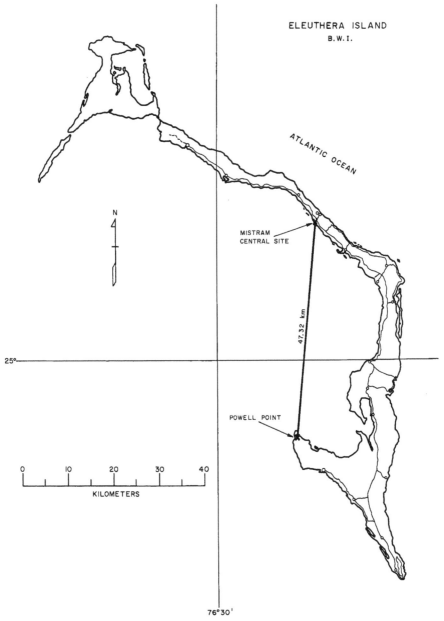

ELEUTHERA ISLAND
B.W.I.

ATLANTIC OCEAN

N

MISTRAM
CENTRAL SITE

47.32 km

25°

POWELL POINT

0   10   20   30   40
KILOMETERS

76°30'

Figure 1

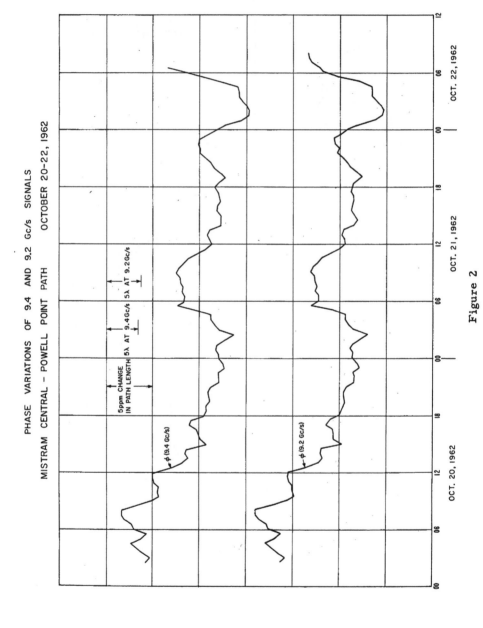

PHASE VARIATIONS OF 9.4 AND 9.2 Gc/s SIGNALS

MISTRAM CENTRAL - POWELL POINT PATH   OCTOBER 20-22, 1962

Figure 2

-20-

PHASE VARIATIONS OF
9.4 AND 9.2 GC/S SIGNALS
MISTRAM CENTRAL – POWELL POINT PATH   OCT 30-31, 1962

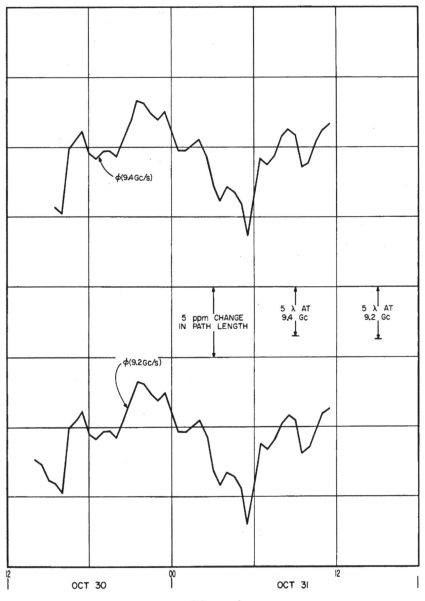

Figure 3

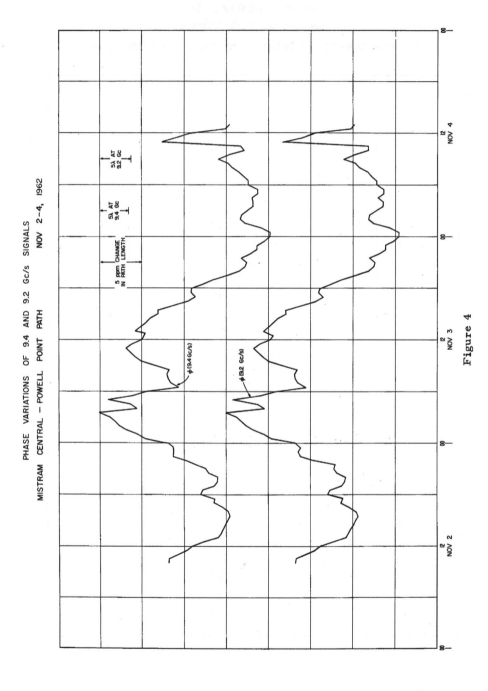

PHASE VARIATIONS OF 9.4 AND 9.2 Gc/s SIGNALS
MISTRAM CENTRAL — POWELL POINT PATH   NOV 2-4, 1962

Figure 4

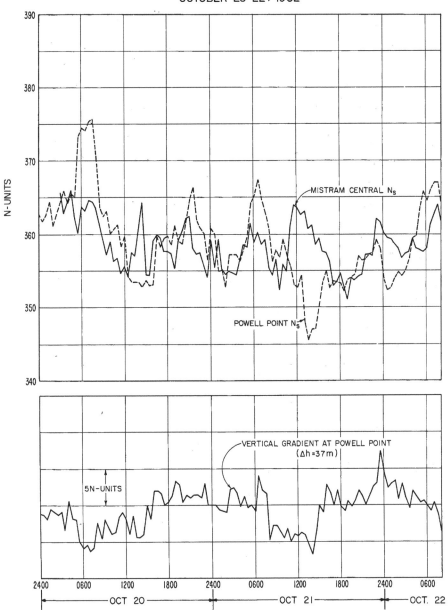

REFRACTIVITY VARIATIONS
OCTOBER 20-22, 1962

Figure 5

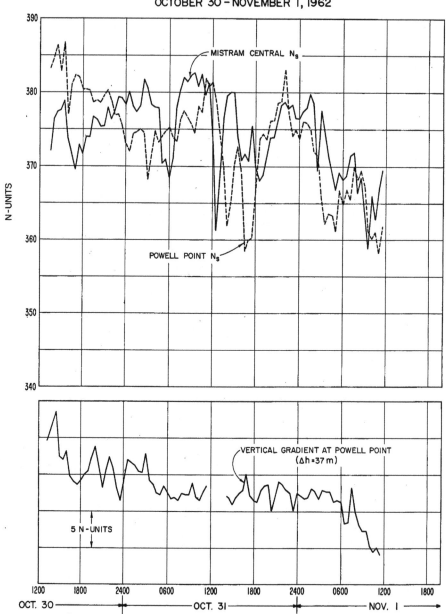

REFRACTIVITY VARIATIONS
OCTOBER 30 – NOVEMBER 1, 1962

Figure 6

-24-

REFRACTIVITY VARIATIONS
NOVEMBER 2-4, 1962

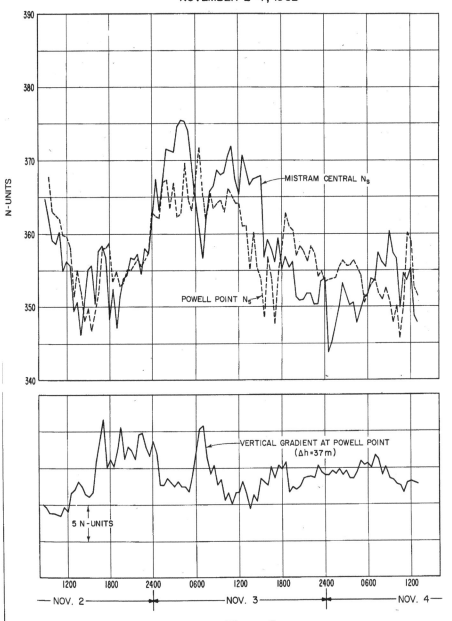

Figure 7

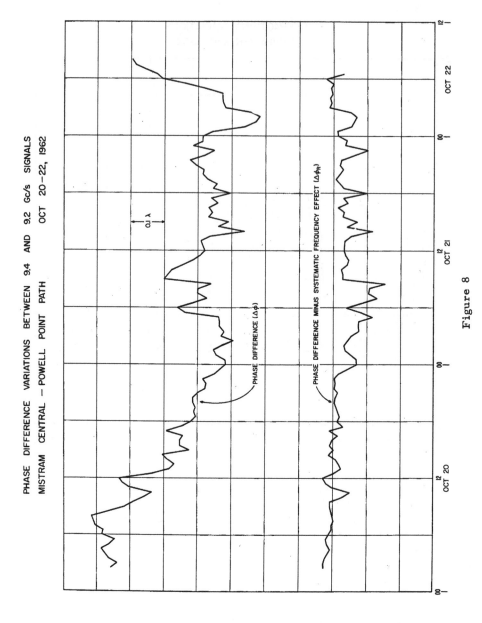

PHASE DIFFERENCE VARIATIONS BETWEEN 9.4 AND 9.2 Gc/s SIGNALS
MISTRAM CENTRAL — POWELL POINT PATH   OCT 20−22, 1962

Figure 8

-26-

PHASE DIFFERENCE VARIATIONS BETWEEN
9.4 AND 9.2 GC/S SIGNALS
MISTRAM CENTRAL – POWELL POINT PATH   OCT 30-31, 1962

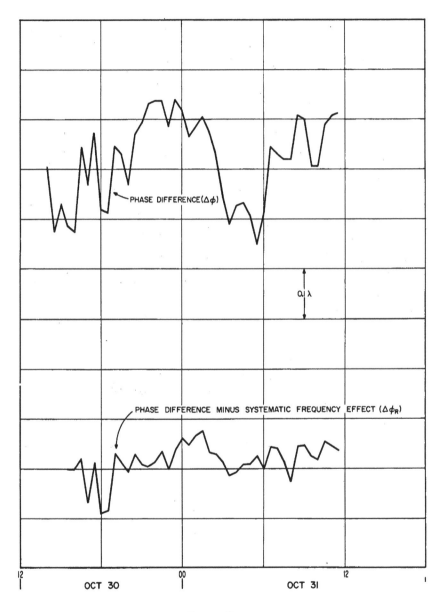

Figure 9

-27-

PHASE DIFFERENCE VARIATIONS BETWEEN 9.4 AND 9.2 Gc/s SIGNALS
MISTRAM CENTRAL — POWELL POINT PATH   NOV 2-4, 1962

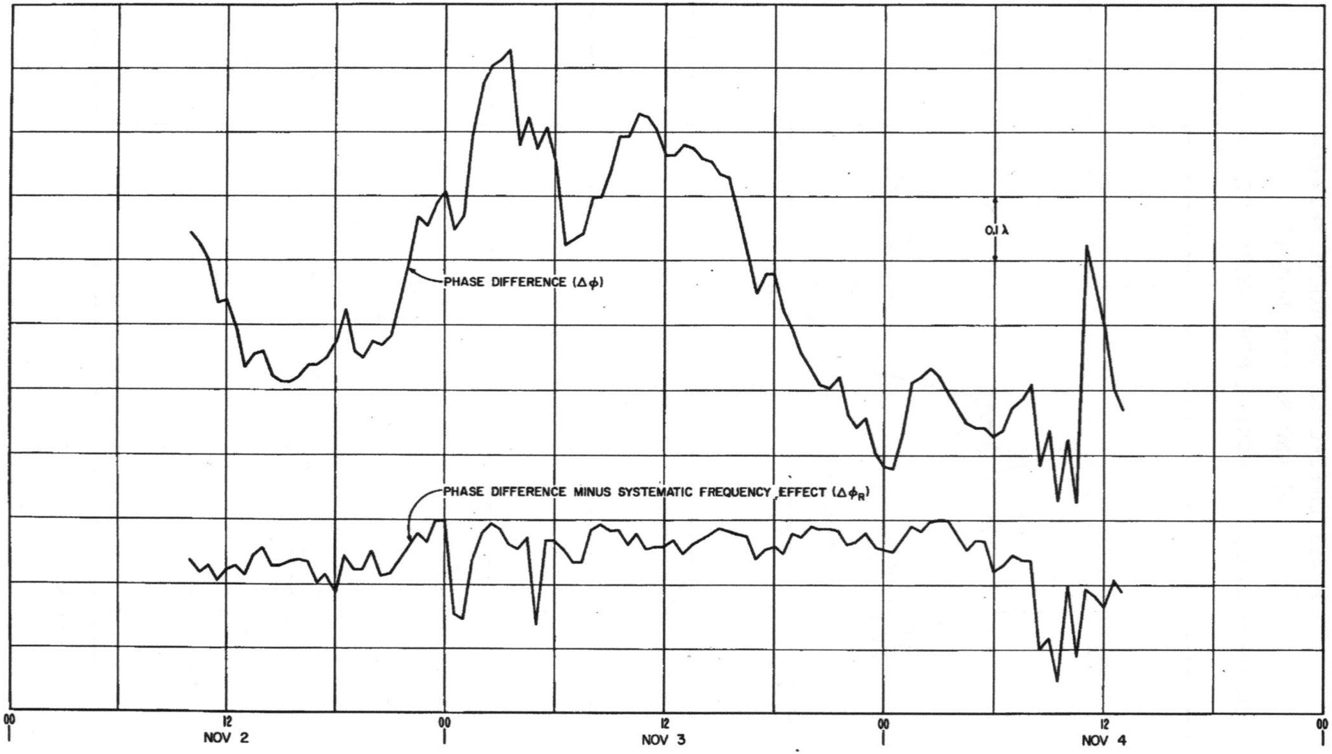

Figure 10

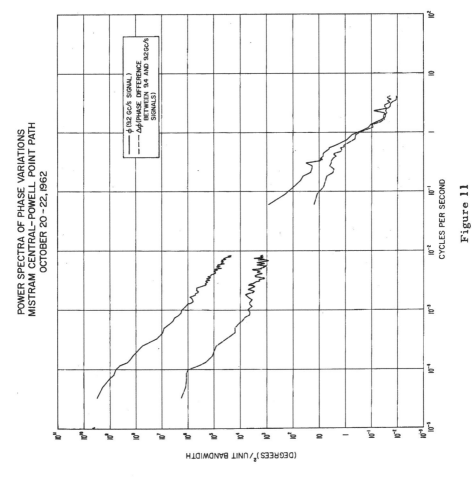

POWER SPECTRA OF PHASE VARIATIONS
MISTRAM CENTRAL-POWELL POINT PATH
OCTOBER 20-22, 1962

ϕ (9.2 GC/s SIGNAL)
$\Delta\phi$ (PHASE DIFFERENCE BETWEEN 9.4 AND 9.2 GC/s SIGNALS)

CYCLES PER SECOND

(DEGREES)^2/UNIT BANDWIDTH

Figure 11

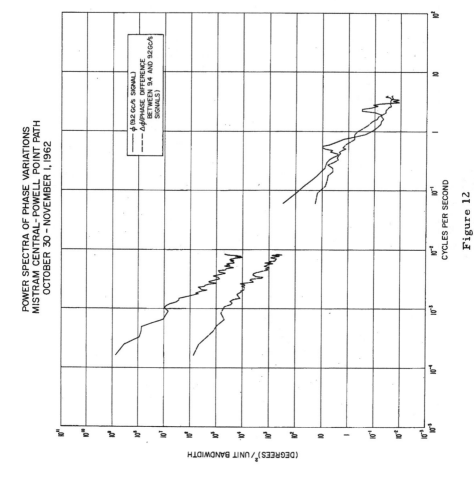

Figure 12

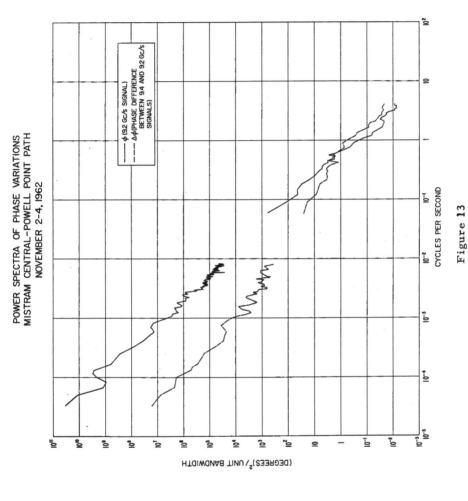

POWER SPECTRA OF PHASE VARIATIONS
MISTRAM CENTRAL-POWELL POINT PATH
NOVEMBER 2-4, 1962

φ (9.2 Gc/s SIGNAL)
Δφ(PHASE DIFFERENCE
BETWEEN 9.4 AND 9.2 Gc/s
SIGNALS)

CYCLES PER SECOND

Figure 13

-31-

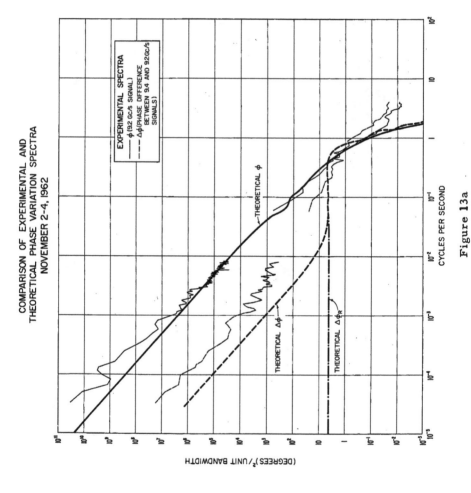

COMPARISON OF EXPERIMENTAL AND
THEORETICAL PHASE VARIATION SPECTRA
NOVEMBER 2-4, 1962

EXPERIMENTAL SPECTRA
——— φ (9.2 GC/S SIGNAL)
--- Δφ(PHASE DIFFERENCE
     BETWEEN 9.4 AND 9.2GC/S
     SIGNALS)

THEORETICAL φ

THEORETICAL Δφ

THEORETICAL Δφ$_R$

(DEGREES)^2/UNIT BANDWIDTH

CYCLES PER SECOND

Figure 13a

-32-

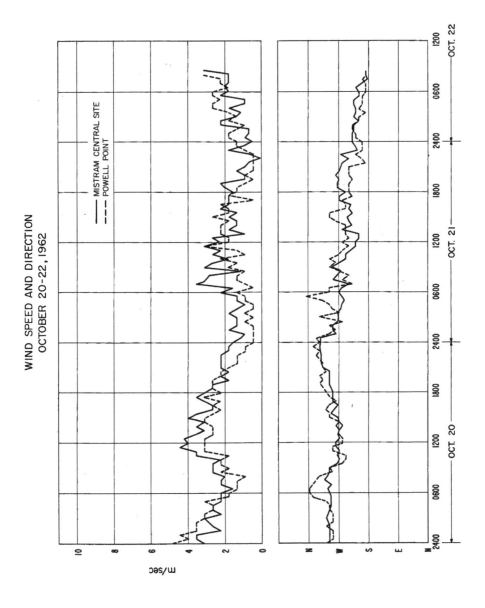

Figure 14

-33-

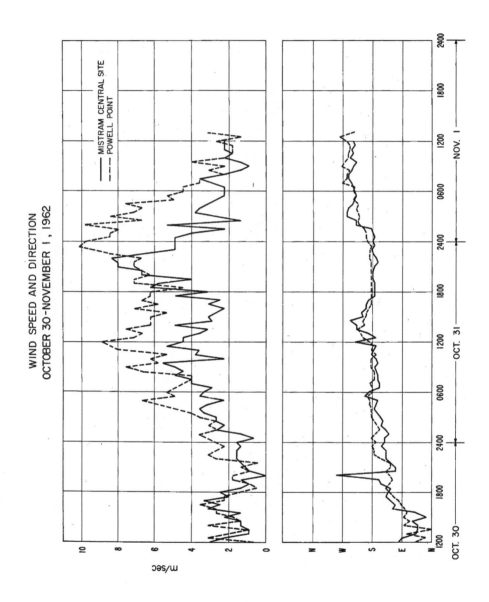

WIND SPEED AND DIRECTION
OCTOBER 30-NOVEMBER 1, 1962

MISTRAM CENTRAL SITE
POWELL POINT

Figure 15

-34-

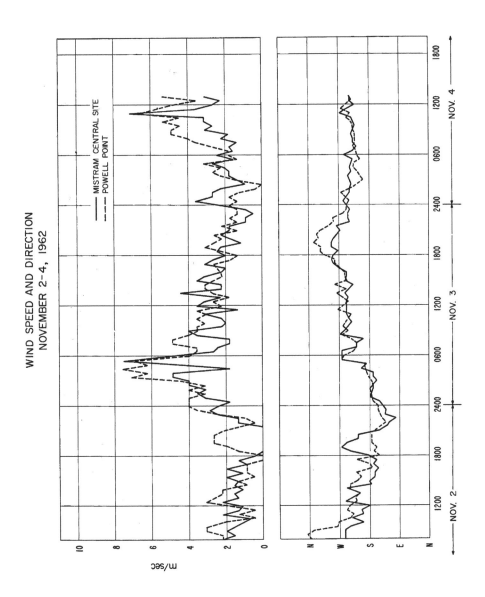

Figure 16

CUMULATIVE DISTRIBUTIONS OF 9.4 AND 9.2 Gc/s
PHASE DIFFERENCE VARIATIONS
MISTRAM CENTRAL - POWELL POINT PATH
OCTOBER 20-22, 1962
(3-MIN. SAMPLES; PASSBAND 004-15 c/s)

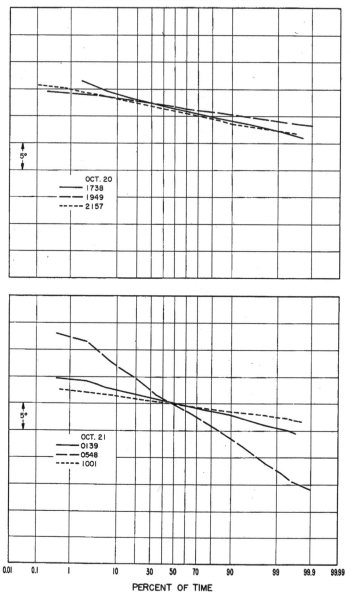

Figure 17

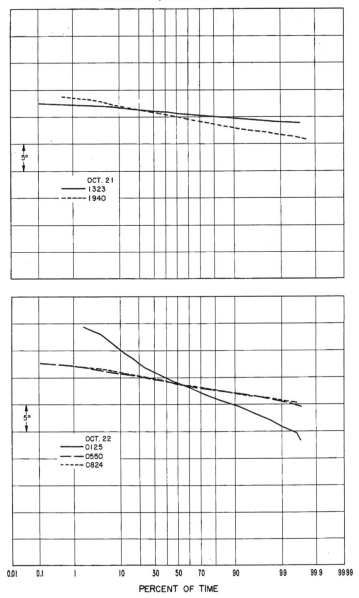

CUMULATIVE DISTRIBUTIONS OF 9.4 AND 9.2 Gc/s
PHASE DIFFERENCE VARIATIONS
MISTRAM CENTRAL - POWELL POINT PATH
OCTOBER 20-22, 1962
(3-MIN. SAMPLES, PASSBAND: 004-15 c/s)

Figure 18

CUMULATIVE DISTRIBUTIONS OF 94 AND 92 Gc/s
PHASE DIFFERENCE VARIATIONS
MISTRAM CENTRAL - POWELL POINT PATH
OCTOBER 30 - NOVEMBER 1, 1962
(3-MIN. SAMPLES; PASSBAND 0.04-15 c/s)

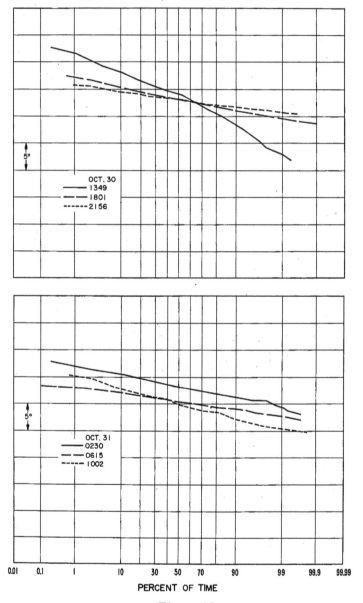

PERCENT OF TIME

Figure 19

CUMULATIVE DISTRIBUTIONS OF 9.4 AND 9.2 Gc/s
PHASE DIFFERENCE VARIATIONS
MISTRAM CENTRAL - POWELL POINT PATH
OCTOBER 30 - NOVEMBER 1, 1962
(3 - MIN. SAMPLES; PASSBAND: 0.04 - 15 c/s)

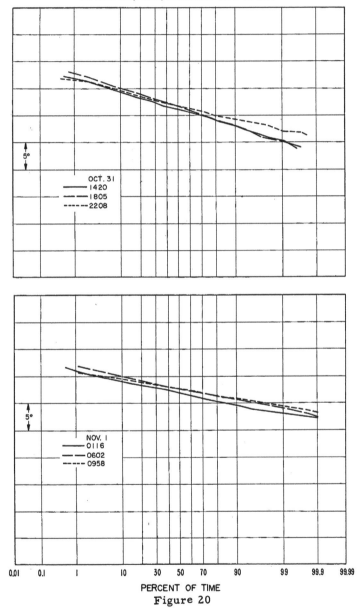

Figure 20

CUMULATIVE DISTRIBUTIONS OF 9.4 AND 9.2 Gc/s
PHASE DIFFERENCE VARIATIONS
MISTRAM CENTRAL - POWELL POINT PATH
NOVEMBER 2-4, 1962
(3-MIN. SAMPLES; PASSBAND: 0.04-15 c/s)

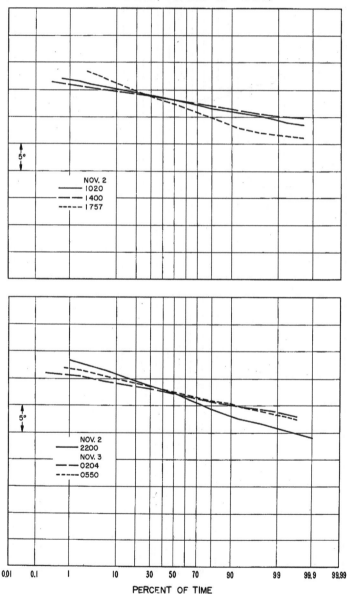

PERCENT OF TIME

Figure 21

CUMULATIVE DISTRIBUTIONS OF 9.4 AND 9.2 Gc/s
PHASE DIFFERENCE VARIATIONS
MISTRAM CENTRAL - POWELL POINT PATH
NOVEMBER 2-4, 1962
(3-MIN. SAMPLES; PASSBAND: 0.04-15 c/s)

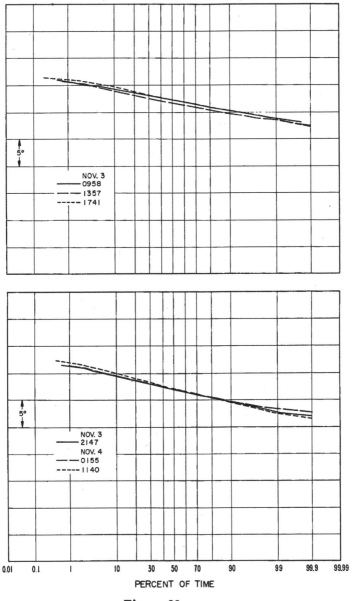

Figure 22

-41-

VARIATIONS IN PHASE DIFFERENCE OF 9.4 AND 9.2 GC/s SIGNALS

MISTRAM CENTRAL – POWELL POINT PATH

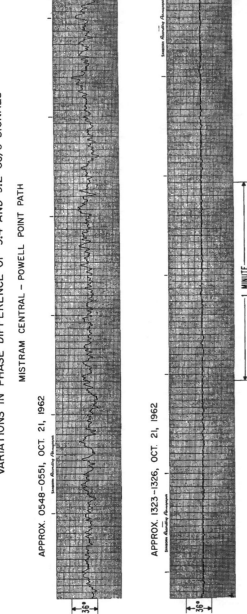

APPROX. 0548-0551, OCT. 21, 1962

APPROX. 1323-1326, OCT. 21, 1962

Figure 23

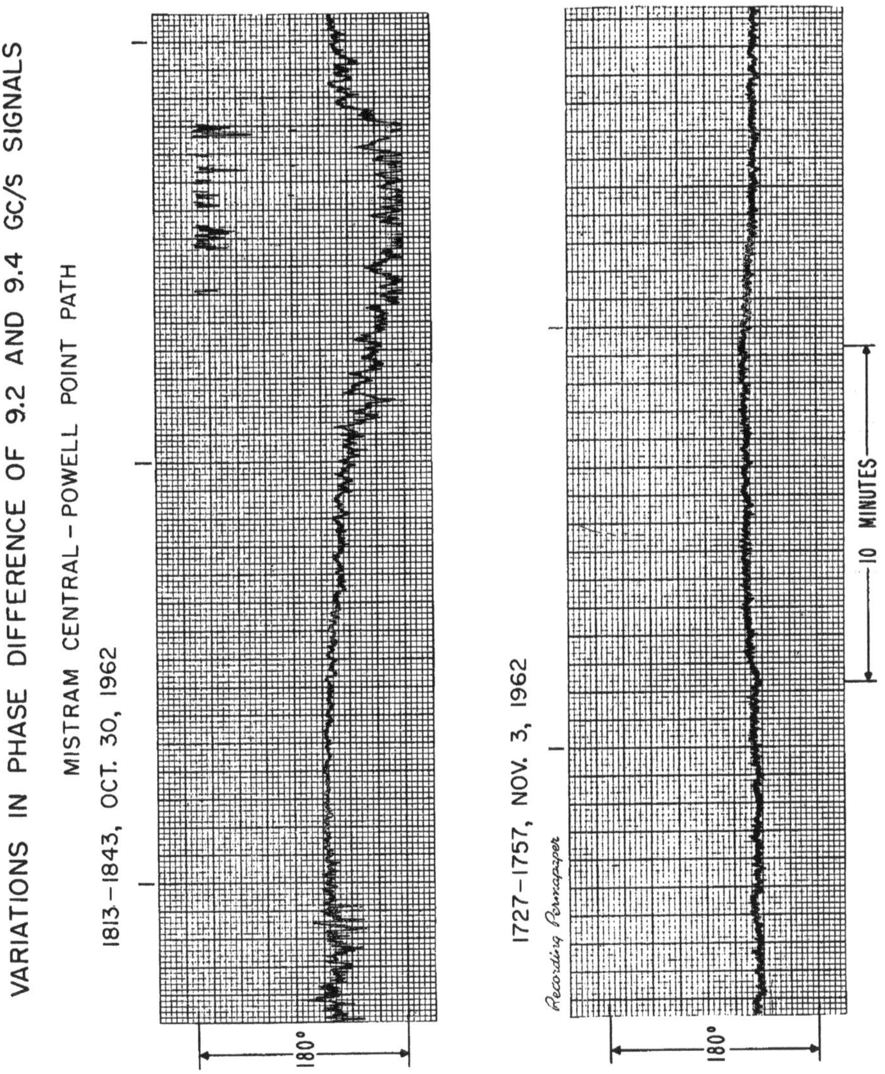

VARIATIONS IN PHASE DIFFERENCE OF 9.2 AND 9.4 GC/S SIGNALS

MISTRAM CENTRAL – POWELL POINT PATH

1813 – 1843, OCT. 30, 1962

1727 – 1757, NOV. 3, 1962

Figure 24

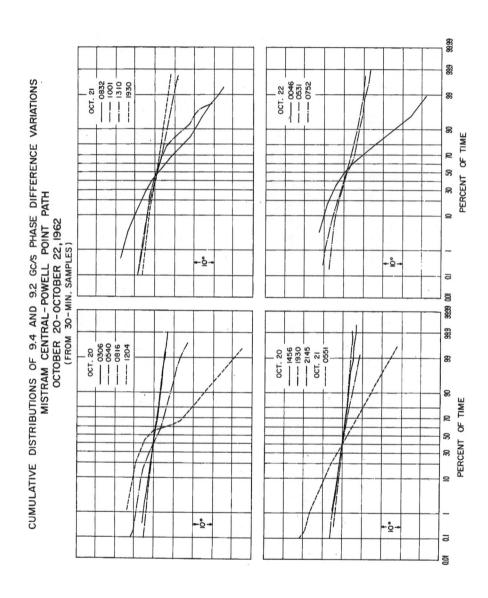

CUMULATIVE DISTRIBUTIONS OF 9.4 AND 9.2 GC/S PHASE DIFFERENCE VARIATIONS
MISTRAM CENTRAL-POWELL POINT PATH
OCTOBER 20-OCTOBER 22, 1962
(FROM 30-MIN. SAMPLES)

Figure 25

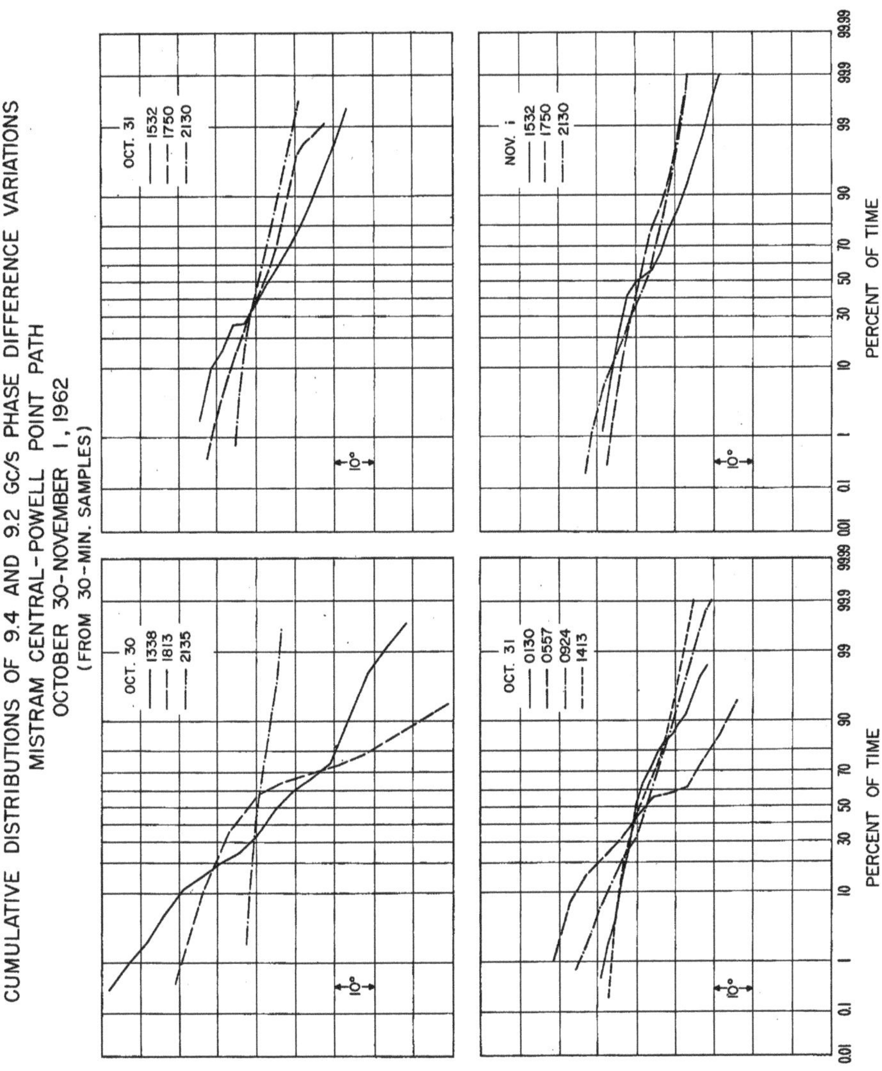
CUMULATIVE DISTRIBUTIONS OF 9.4 AND 9.2 GC/S PHASE DIFFERENCE VARIATIONS
MISTRAM CENTRAL-POWELL POINT PATH
OCTOBER 30-NOVEMBER 1, 1962
(FROM 30-MIN. SAMPLES)
OCT. 30
1338
1813
2135
OCT. 31
1532
1750
2130
OCT. 31
0130
0557
0924
1413
NOV. 1
1532
1750
2130
10°
0.01
0.1
1
10
30
50
70
90
99
99.9
99.99
PERCENT OF TIME

Figure 26

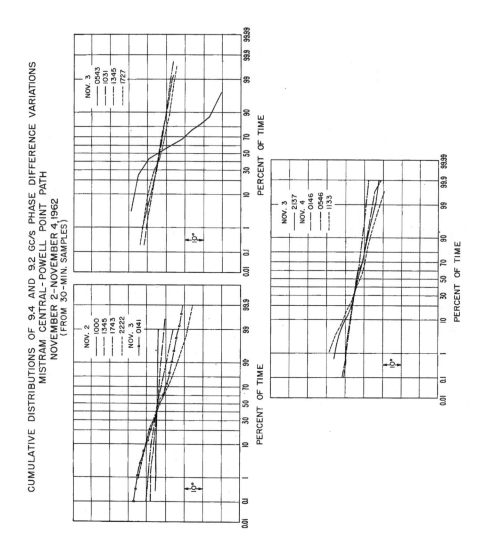

Figure 27

-46-

EXAMPLES OF AMPLITUDE AND DIFFERENTIAL AMPLITUDE VARIATIONS

MISTRAM CENTRAL—POWELL POINT PATH
UPPER CURVES=DIFFERENTIAL AMPLITUDE (9.2-9.4 Gc/s)
LOWER CURVES=AMPLITUDE AT 9.4 Gc/s

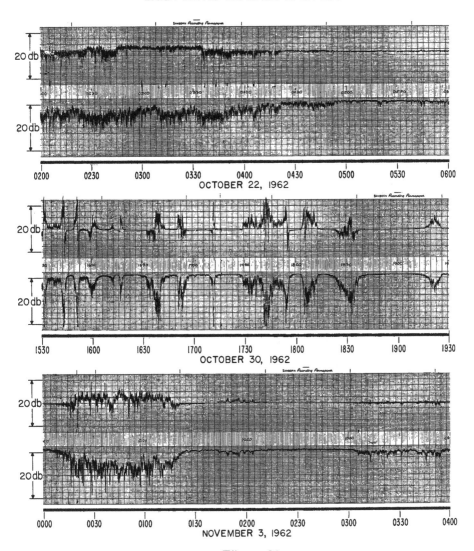

Figure 28

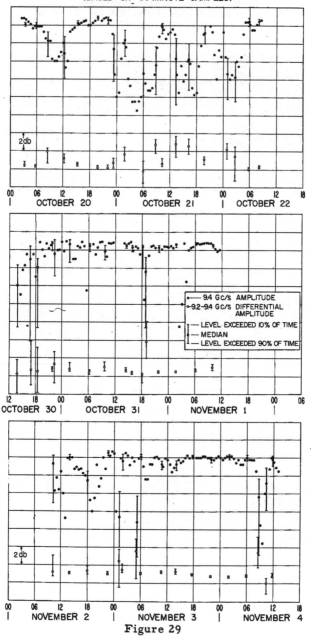

SIGNAL AMPLITUDE VARIATIONS
MISTRAM CENTRAL - POWELL POINT PATH 1962
(BASED ON 30 MINUTE SAMPLES)

Figure 29

-48-

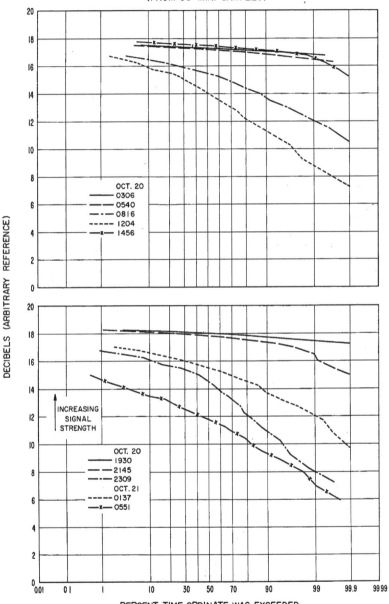

CUMULATIVE DISTRIBUTIONS OF 9.4 Gc/s SIGNAL AMPLITUDE
MISTRAM CENTRAL – POWELL POINT PATH
OCTOBER 20-22, 1962
(FROM 30-MIN. SAMPLES)

Figure 30

-49-

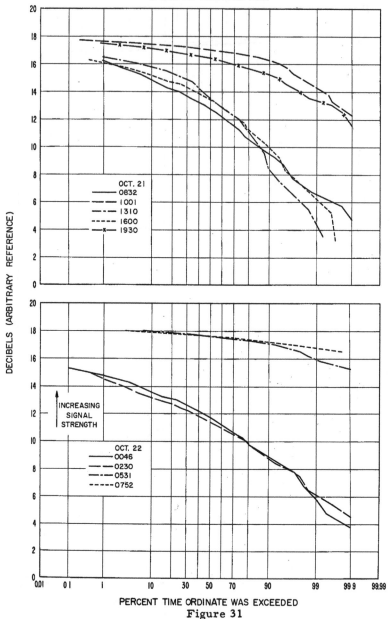

CUMULATIVE DISTRIBUTIONS OF 9.4 Gc/s SIGNAL AMPLITUDE
MISTRAM CENTRAL - POWELL POINT PATH
OCTOBER 20-22, 1962
(FROM 30-MIN. SAMPLES)

Figure 31

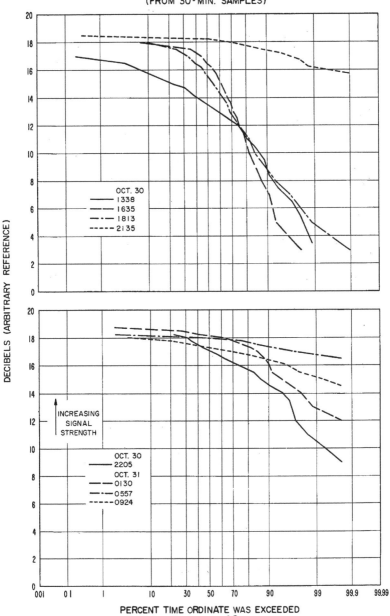

CUMULATIVE DISTRIBUTIONS OF 9.4 Gc/s SIGNAL AMPLITUDE
MISTRAM CENTRAL - POWELL POINT PATH
OCTOBER 30 - NOVEMBER 1, 1962
(FROM 30 - MIN. SAMPLES)

OCT. 30
——— 1338
——— 1635
—·—· 1813
----- 2135

INCREASING
SIGNAL
STRENGTH

OCT. 30
——— 2205
OCT. 31
——— 0130
—·—· 0557
----- 0924

DECIBELS (ARBITRARY REFERENCE)

PERCENT TIME ORDINATE WAS EXCEEDED

Figure 32

-51-

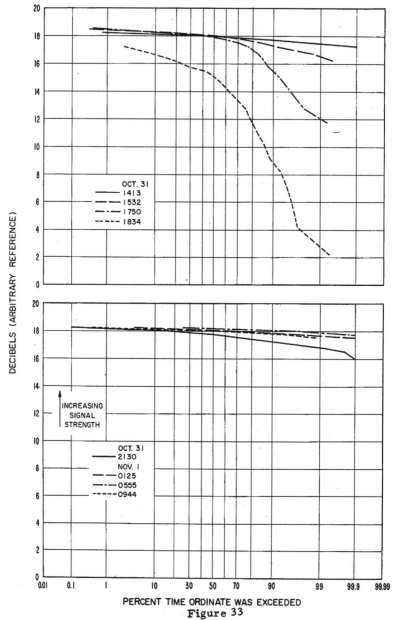

CUMULATIVE DISTRIBUTIONS OF 9.4 Gc/s SIGNAL AMPLITUDE
MISTRAM CENTRAL - POWELL POINT PATH
OCTOBER 30 - NOVEMBER 1, 1962
(FROM 30-MIN. SAMPLES)

Figure 33

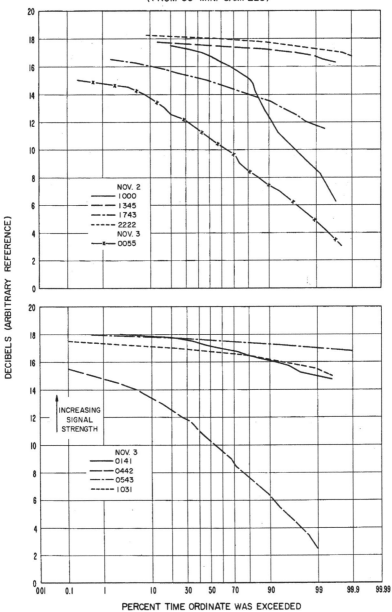

CUMULATIVE DISTRIBUTIONS OF 9.4 Gc/s SIGNAL AMPLITUDE
MISTRAM CENTRAL -POWELL POINT PATH
NOVEMBER 2-4, 1962
(FROM 30-MIN. SAMPLES)

Figure 34

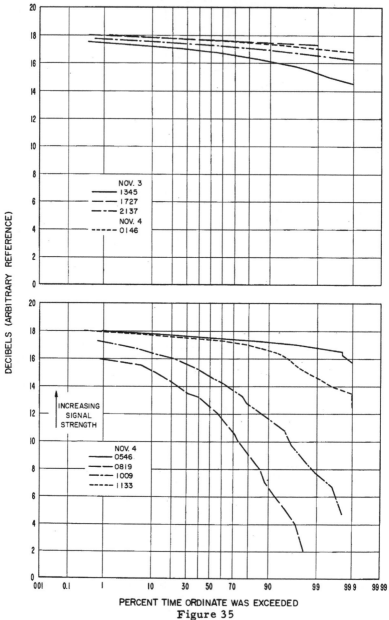

CUMULATIVE DISTRIBUTIONS OF 9.4 Gc/s SIGNAL AMPLITUDE
MISTRAM CENTRAL - POWELL POINT PATH
NOVEMBER 2-4, 1962
(FROM 30-MIN. SAMPLES)

NOV. 3
——— 1345
——— 1727
——·— 2137
NOV. 4
----- 0146

DECIBELS (ARBITRARY REFERENCE)

INCREASING
SIGNAL
STRENGTH

NOV. 4
——— 0546
——— 0819
——·— 1009
----- 1133

PERCENT TIME ORDINATE WAS EXCEEDED

Figure 35

-54-

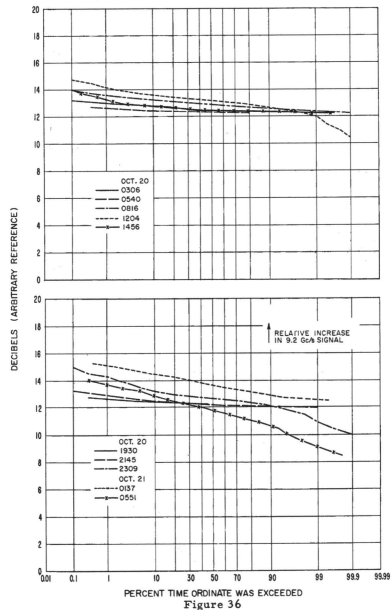

CUMULATIVE DISTRIBUTIONS OF 9.4 AND 9.2 Gc/s
DIFFERENTIAL AMPLITUDE
MISTRAM CENTRAL - POWELL POINT PATH
OCTOBER 20-22, 1962
(FROM 30-MIN. SAMPLES STARTING AT TIME NOTED)

Figure 36

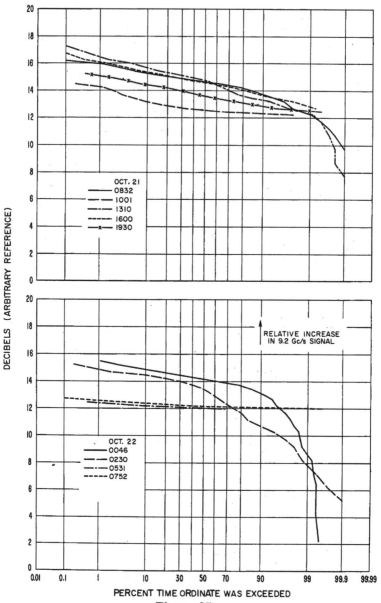

CUMULATIVE DISTRIBUTIONS OF 9.4 AND 9.2 Gc/s
DIFFERENTIAL AMPLITUDE
MISTRAM CENTRAL – POWELL POINT PATH
OCTOBER 20-22, 1962
(FROM 30-MIN. SAMPLES STARTING AT TIME NOTED)

Figure 37

-56-

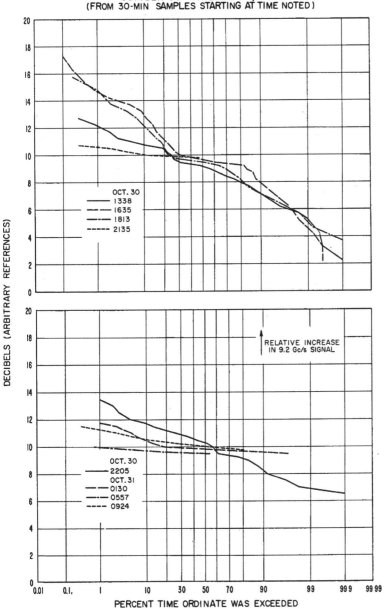

CUMULATIVE DISTRIBUTIONS OF 9.4 AND 9.2 Gc/s
DIFFERENTIAL AMPLITUDE
MISTRAM CENTRAL - POWELL POINT PATH
OCTOBER 30 - NOVEMBER 1, 1962
(FROM 30-MIN SAMPLES STARTING AT TIME NOTED)

OCT. 30
——— 1338
——— 1635
—·—· 1813
------ 2135

RELATIVE INCREASE
IN 9.2 Gc/s SIGNAL

OCT. 30
——— 2205
OCT. 31
——— 0130
—·—· 0557
------ 0924

DECIBELS (ARBITRARY REFERENCES)

PERCENT TIME ORDINATE WAS EXCEEDED

Figure 38

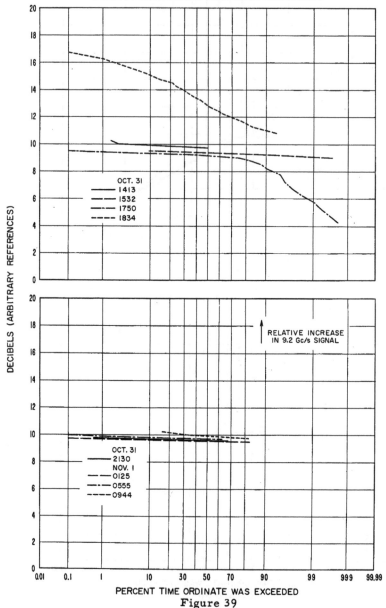

CUMULATIVE DISTRIBUTIONS OF 9.4 AND 9.2 Gc/s
DIFFERENTIAL AMPLITUDE
MISTRAM CENTRAL - POWELL POINT PATH
OCTOBER 30-NOVEMBER 1, 1962
(FROM 30-MIN. SAMPLES STARTING AT TIME NOTED)

Figure 39

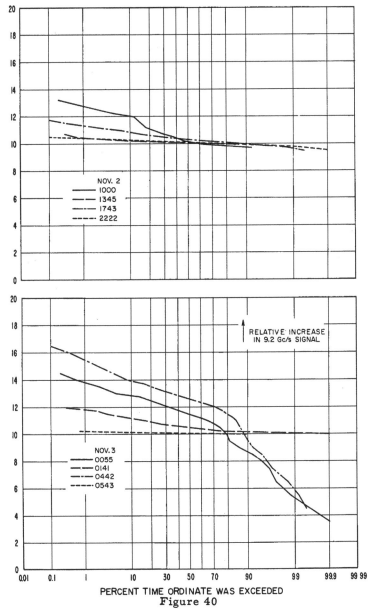

CUMULATIVE DISTRIBUTIONS OF 9.4 AND 9.2 Gc/s
DIFFERENTIAL AMPLITUDE
MISTRAM CENTRAL - POWELL POINT PATH
NOVEMBER 2-4, 1962
(FROM 30-MIN. SAMPLES STARTING AT TIME NOTED)

NOV. 2
——— 1000
——— 1345
——·— 1743
——··— 2222

RELATIVE INCREASE
IN 9.2 Gc/s SIGNAL

NOV. 3
——— 0055
——— 0141
——·— 0442
——··— 0543

DECIBELS (ARBITRARY REFERENCES)

PERCENT TIME ORDINATE WAS EXCEEDED

Figure 40

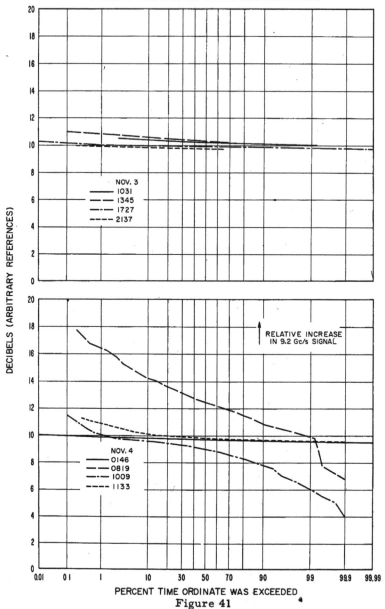

CUMULATIVE DISTRIBUTIONS OF 9.4 AND 9.2 Gc/s
DIFFERENTIAL AMPLITUDE
MISTRAM CENTRAL - POWELL POINT PATH
NOVEMBER 2-4, 1962
(FROM 30-MIN. SAMPLES STARTING AT TIME NOTED)

Figure 41

-60-

VARIATIONS IN DIFFERENTIAL AMPLITUDE OF 9.2 AND 9.4 GC/S SIGNALS

MISTRAM CENTRAL — POWELL POINT PATH

0230-0300, OCT. 22, 1962

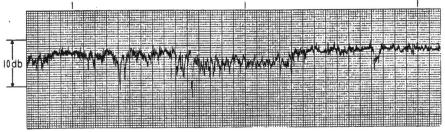

10 db

1813-1843, OCT. 30, 1962

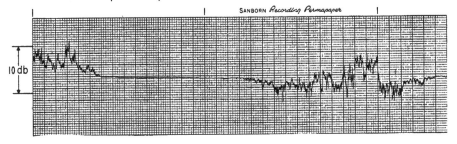

SANBORN *Recording Permapaper*

10 db

1727-1757, NOV. 3, 1962

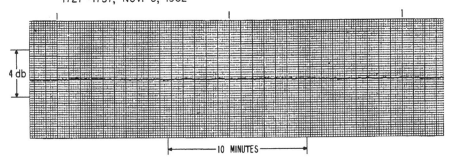

4 db

|← 10 MINUTES →|

Figure 42

Lightning Source UK Ltd.
Milton Keynes UK
UKHW020625060119
334855UK00006B/339/P

9 780265 857977